**Houghton
Mifflin
Harcourt**

© Houghton Mifflin Harcourt Publishing Company • Cover Image Credits: (Grey Wolf pup) ©Don Johnson/All Canada Photos/Getty Images; (Rocky Mountains, Montana) ©Sankar Salvady/Flickr/Getty Images

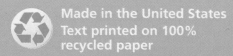

Made in the United States
Text printed on 100%
recycled paper

Houghton Mifflin Harcourt

GO MATH!

ISBN 978-0-544-34190-6

16   0928   19

4500746728   C D E F G

Dear Students and Families,

Welcome to **Go Math!**, Grade 1! In this exciting mathematics program, there are hands-on activities to do and real-world problems to solve. Best of all, you will write your ideas and answers right in your book. In **Go Math!**, writing and drawing on the pages helps you think deeply about what you are learning, and you will really understand math!

By the way, all of the pages in your **Go Math!** book are made using recycled paper. We wanted you to know that you can Go Green with **Go Math!**

Sincerely,

The Authors

Made in the United States
Text printed on 100% recycled paper

GO MATH!

Authors

Juli K. Dixon, Ph.D.
Professor, Mathematics Education
University of Central Florida
Orlando, Florida

Edward B. Burger, Ph.D.
President, Southwestern University
Georgetown, Texas

Steven J. Leinwand
Principal Research Analyst
American Institutes for
    Research (AIR)
Washington, D.C.

Contributor

Rena Petrello
Professor, Mathematics
Moorpark College
Moorpark, California

Matthew R. Larson, Ph.D.
K-12 Curriculum Specialist for
    Mathematics
Lincoln Public Schools
Lincoln, Nebraska

Martha E. Sandoval-Martinez
Math Instructor
El Camino College
Torrance, California

English Language Learners Consultant

Elizabeth Jiménez
CEO, GEMAS Consulting
Professional Expert on English
    Learner Education
Bilingual Education and
    Dual Language
Pomona, California

Number and Operations in Base Ten

Critical Area Developing understanding of whole number relationships and place value, including grouping in tens and ones

Vocabulary Reader **Around the Neighborhood** **319**

6 Count and Model Numbers 327

COMMON CORE STATE STANDARDS

1.NBT Number and Operations in Base Ten
Cluster A Extend the counting sequence.
1.NBT.A.1
Cluster B Understand place value.
1.NBT.B.2
1.NBT.B.2a
1.NBT.B.2b
1.NBT.B.2c
1.NBT.B.3

☑ **Show What You Know** **328**

Vocabulary Builder **329**

Game: Show the Numbers **330**

Chapter Vocabulary Cards

Vocabulary Game **330A**

1 **Count by Ones to 120** **331**
   Practice and Homework

2 **Count by Tens to 120** **337**
   Practice and Homework

3 **Understand Ten and Ones** **343**
   Practice and Homework

4 **Hands On • Make Ten and Ones** **349**
   Practice and Homework

5 **Hands On • Tens** **355**
   Practice and Homework

☑ **Mid-Chapter Checkpoint** **358**

GO DIGITAL

Go online! Your math lessons are interactive. Use *i*Tools, Animated Math Models, the Multimedia *e*Glossary, and more.

Essential Question
How can knowing a counting pattern help you count to 120?
Start

Chapter 6 Overview

In this chapter, you will explore and discover answers to the following **Essential Questions**:

- How do you use place value to model, read, and write numbers to 120?
- What ways can you use tens and ones to model numbers to 120?
- How do numbers change as you count by tens to 120?

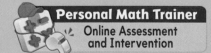

Personal Math Trainer
Online Assessment and Intervention

6 Hands On • **Tens and Ones to 50** 361
Practice and Homework

7 Hands On • **Tens and Ones to 100** 367
Practice and Homework

8 **Problem Solving • Show Numbers in Different Ways** . . . 373
Practice and Homework

9 Hands On • **Model, Read, and Write Numbers**
from 100 to 110 379
Practice and Homework

10 Hands On • **Model, Read, and Write Numbers**
from 110 to 120 385
Practice and Homework

✔ **Chapter 6 Review/Test** 391

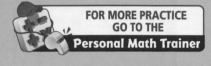

FOR MORE PRACTICE
GO TO THE
Personal Math Trainer

Practice and Homework

Lesson Check and Spiral Review in every lesson

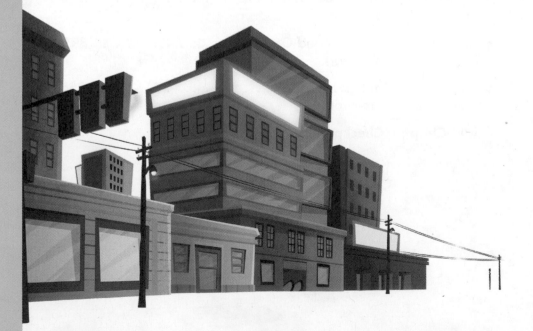

Around the Neighbor-hood

written by John Hudson

UNITED STATES POST OFFICE

We Deliver

Common Core

CRITICAL AREA Developing understanding of whole number relationships and place value, including grouping in tens and ones

The mail carrier brings letters to Mr. and Mrs. Jones. How many letters does she bring?

_____ ◯ _____ ◯ _____

Social Studies

320

How do mail carriers help us?

The mail carrier brings packages to the fire station. Then she brings more packages. How many packages does she bring?

_____ ◯ _____ ◯ _____

How do firefighters help us?

It is time for lunch. The mail carrier eats in the park. How many boys and girls are playing?

_____ ◯ _____ ◯ _____

Social Studies

How do parents help us?

The mail carrier brings 12 packages to the police station. "This person has moved," says the police officer. "You need to take these back." How many packages does the officer keep?

_____ ◯ _____ ◯ _____

Social Studies

How do police officers help us?

The mail carrier stops at City Hall.
She brings 8 letters for the mayor.
She brings 4 letters for the city clerk.
How many letters does she bring?

_____ ◯ _____ ◯ _____

Social Studies

How do city workers help us?

Write About the Story

One day, Mr. and Mrs. Jones each got the same number of letters. They got 12 letters in all. Draw the two groups of letters.

Vocabulary Review

add   difference

doubles   subtract

sum

Mr. Jones

Mrs. Jones

Letters for Mr. Jones

Letters for Mrs. Jones

Write the number sentence. _____ ◯ _____ ◯ _____

 Describe your number sentence. Use a vocabulary word.

_____

_____

_____

_____

_____

How Many Letters?

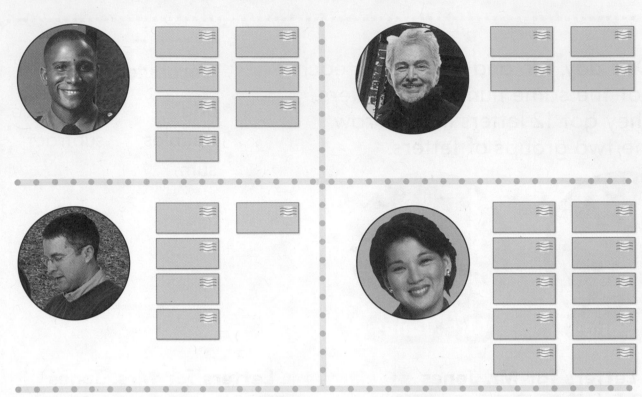

1. How many letters do and have in all?

   _____ ◯ _____ ◯ _____

2. How many more letters does have than ?

   _____ ◯ _____ ◯ _____

3. Circle the two that have 11 letters in all.

MATH BOARD Make up an addition story about the mail carrier bringing letters to you and a classmate. Write the number sentence.

Count and Model Numbers

Curious About Math with Curious George

Dan and May like apples. They buy 15 apples in all. If Dan buys 10 apples, how many apples does May buy?

Name _____

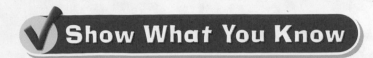

Explore Numbers 6 to 9

Count how many. Circle the number. (K.CC.B.4)

1.

6

7

2.

8

9

Count Groups to 20

Circle groups of 10. Write how many. (1.NBT.A.1)

3.

_____

4.

_____

Make Groups of 10

Use ●. Draw to show a group of 10 in two different ways. (1.NBT.B.2a)

5.

6.

This page checks understanding of important skills needed for success in Chapter 6.

Vocabulary Builder

Review Words

one	two
three	four
five	six
seven	eight
nine	ten

Visualize It
Draw pictures in the box to show the number.

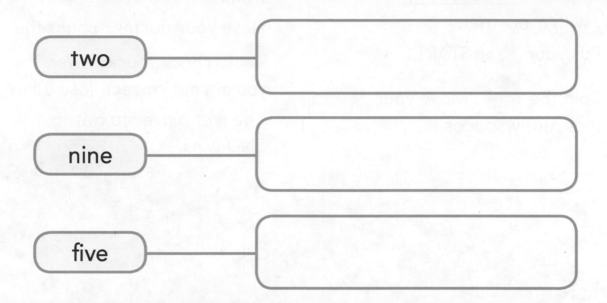

two

nine

five

Understand Vocabulary
Write a review word to name the number.

1.  _____

2. _____

3. _____

GO
DIGITAL
• Interactive Student Edition
• Multimedia eGlossary

Game Show the Numbers

Materials • and •
• 20 ⬤ • ⬜⬜⬜⬜⬜ / ⬜⬜⬜⬜⬜

Play with a partner.

1. Put your on START.

2. Spin the . Move your that many spaces.

3. Read the number. Use ⬤ to show the number on a ten frame.

4. Have your partner count the ⬤ to check your answer. If you are not correct, lose a turn.

5. The first player to get to END wins.

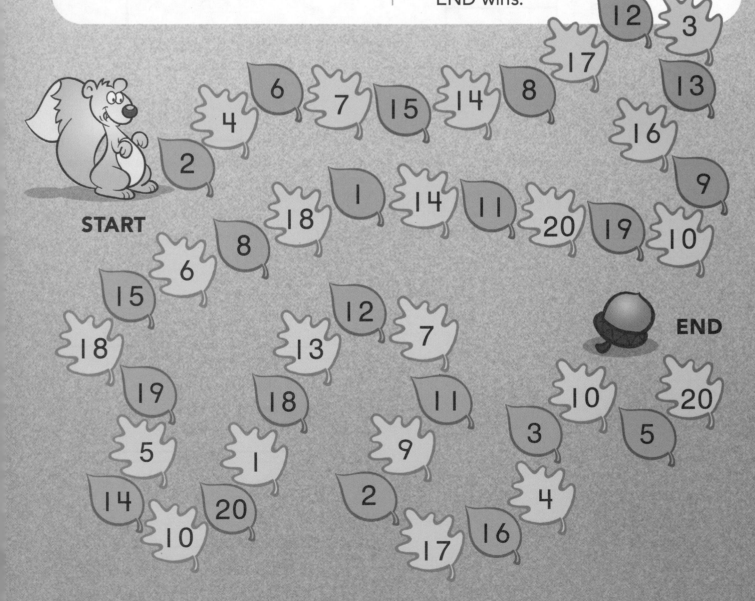

Chapter 6 Vocabulary

addend

sumando

2

difference

diferencia

13

digit

dígito

14

hundred

centena

29

ones

unidades

39

subtract

restar

52

sum

suma o total

54

ten

decena

57

$9 - 4 = 5$

The **difference** is 5.

5   +   3   =   8

↑

addends

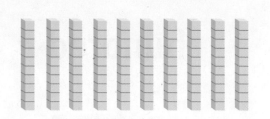

10 tens is the same as **1 hundred**.

13 is a two-digit number.

The 1 in 13 means 1 ten.

The 3 in 13 means 3 ones.

$5 - 2 = 3$

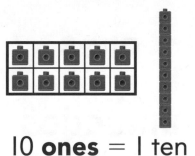

10 **ones** = 1 ten

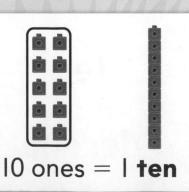

10 ones = 1 **ten**

2 plus 1 is equal to 3.

The **sum** is 3.

Going to Town

For 2 players

Materials

- I 🔲
- I 🔲
- I 🎲
- clue cards

How to Play

1. Put your 🔲 on START.
2. Toss the 🎲 to take a turn. Move that many spaces.
3. If you land on one of these spaces:

   Red Space Take a Clue Card. Answer the question. If you are correct, move ahead I. Return the Clue Card to the bottom of the pile.

   Blue Space Read the word. Tell what it means, or give an example. If you are correct, move ahead I.

   Green Space Follow the directions in the space.
4. The first player to reach FINISH wins.

Word Box

addend

difference

digit

hundred

ones

subtract

sum

ten

Game

DIRECTIONS 2 players. Take turns to play. • To take a turn, toss the 🎲. Move that many spaces • Follow the directions for the space where you land. • First player to reach FINISH wins.

MATERIALS • 1 ⬛ per player • 1 🎲 • 1 set of clue cards

START

CLUE CARD

addend

Play in the park. Move ahead 1.

CLUE CARD

sum

CLUE CARD

Eat at the diner. Lose 1 turn.

digit

FINISH

Shop at the market. Take another turn.

Wait at the post office. Go back **1**.

CLUE CARD

subtract

difference

Go to the library. Trade places with another player.

CLUE CARD

The Write Way

Reflect

Choose one idea. Draw and write about it.

- Tell how you can use a counting chart to solve problems.

- Explain two ways that you can show a number as tens and ones.

Name _____

Count by Ones to 120

Essential Question How can knowing a counting pattern help you count to 120?

Common Core **Number and Operations in Base Ten—1.NBT.A.1**
MATHEMATICAL PRACTICES
MP5, MP7, MP8

Listen and Draw Real World

Write the missing numbers.

21	22	23	24	25	26	27	28	29	30
31	32	33	34	35	36	37	38	39	40
41	42	43	44	45	46	47	48	49	50
51	52	53	54	55	56	57	58	59	60
61	62	63	64	65	66	67	68	69	70
71	72	73	74	75	76	77	78	79	80
81	82	83	84	85	86	87	88	89	90
91	92	93	94	95	96	97	98	99	100

Math Talk MATHEMATICAL PRACTICES 7

Look for Structure Explain how you know which numbers are missing.

FOR THE TEACHER • Read the following problem. Debbie saw this page in a puzzle book. Two rows of numbers are missing. Use what you know about counting to write the missing numbers.

Chapter 6

Count forward.
Write the numbers.

1	2	3	4	5	6	7	8	9	10
11	12	13	14	15	16	17	18	19	20
21	22	23	24	25	26	27	28	29	30
31	32	33	34	35	36	37	38	39	40
41	42	43	44	45	46	47	48	49	50
51	52	53	54	55	56	57	58	59	60
61	62	63	64	65	66	67	68	69	70
71	72	73	74	75	76	77	78	79	80
81	82	83	84	85	86	87	88	89	90
91	92	93	94	95	96	97	98	99	100
101	102	103	104	105	106	107	108	109	110
111	112	113	114	115	116	117	118	119	120

10, _11_, ____, ____, ____

100, _101_, ____, ____, ____

110, _111_, ____, ____, ____

Share and Show   MATH BOARD

Use a Counting Chart. Count forward.
Write the numbers.

Look for a pattern to help you write the numbers.

1. 114, ____, ____, ____, ____, ____

2. 51, ____, ____, ____, ____, ____

3. 94, ____, ____, ____, ____, ____

4. 78, ____, ____, ____, ____, ____

☑5. 35, ____, ____, ____, ____, ____

☑6. 104, ____, ____, ____, ____, ____

Name _____

On Your Own

Look for a Pattern

Use a Counting Chart. Count forward.
Write the numbers.

7. 19, ____, ____, ____, ____, ____, ____, ____

8. 98, ____, ____, ____, ____, ____, ____, ____

9. 60, ____, ____, ____, ____, ____, ____, ____

10. 27, ____, ____, ____, ____, ____, ____, ____

11. 107, ____, ____, ____, ____, ____, ____, ____

12. **THINK SMARTER** Use a Counting Chart to write the numbers counting forward.

____, ____, ____, ____, ____, 120

13. **THINK SMARTER** There is an unknown number in the sequence counting forward. The number is greater than 51. The number is less than 53. What is the unknown number? ____

Problem Solving • Applications Real World   WRITE Math

Use a Counting Chart. Draw and
write numbers to solve.

14. **GO DEEPER** The bag has 99 buttons.
Draw more buttons so there
are 105 buttons in all. Write the
numbers as you count.

99

15. **THINK SMARTER** The bag has
56 buttons. How many more
buttons do you need to add to
the bag to have 64 buttons?

_____ buttons

56

16. **THINK SMARTER** Tito counts 105 cubes. Then he counts
forward some more cubes. Write the numbers.

105, _____, _____, _____, _____, _____, _____

 TAKE HOME ACTIVITY • Take a walk with your child. Count aloud
together as you take 120 steps.

Count by Ones to 120

Common Core **COMMON CORE STANDARD—1.NBT.A.1**
Extend the counting sequence.

Use a Counting Chart. Count forward. Write the numbers.

1. 40, ____, ____, ____, ____, ____, ____, ____, ____

2. 55, ____, ____, ____, ____, ____, ____, ____, ____

3. 37, ____, ____, ____, ____, ____, ____, ____, ____

4. 98, ____, ____, ____, ____, ____, ____, ____, ____

Problem Solving

Use a Counting Chart. Draw and write numbers to solve.

5. The bag has 111 marbles. Draw more marbles so there are 117 marbles in all. Write the numbers as you count.

6. **WRITE ▶ Math** Choose a number between 90 and 110. Write the number. Then count forward to write the next 5 numbers.

_____

_____

_____

1. Count forward. Write the missing number.

110, 111, 112, _____, 114

2. Solve. Write the number. There are 6 bees. 2 bees fly away. How many bees are there now?

_____ bees

3. Solve. Draw a model to explain. There are 8 children. 6 children are boys. The rest are girls. How many children are girls?

_____ girls

© Houghton Mifflin Harcourt Publishing Company

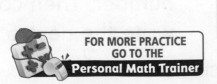

FOR MORE PRACTICE
GO TO THE
Personal Math Trainer

Name _____

Count by Tens to 120

Essential Question How do numbers change as you count by tens to 120?

Common Core — Number and Operations in Base Ten—1.NBT.A.1
MATHEMATICAL PRACTICES
MP2, MP5, MP8

Listen and Draw

Start on 10. Count forward by tens.
Color each number as you say it.

1	2	3	4	5	6	7	8	9	10
11	12	13	14	15	16	17	18	19	20
21	22	23	24	25	26	27	28	29	30
31	32	33	34	35	36	37	38	39	40
41	42	43	44	45	46	47	48	49	50
51	52	53	54	55	56	57	58	59	60
61	62	63	64	65	66	67	68	69	70
71	72	73	74	75	76	77	78	79	80
81	82	83	84	85	86	87	88	89	90
91	92	93	94	95	96	97	98	99	100

Math Talk

MATHEMATICAL PRACTICES 8

Generalize Which numbers in the hundred chart did you color? Explain.

FOR THE TEACHER • Ask children: How do you count by tens? Starting at 10 on the hundred chart, have children count forward by tens, coloring each additional ten as they count.

Model and Draw

Start on 3. Count by tens.

1	2	3	4	5	6	7	8	9	10
11	12	13	14	15	16	17	18	19	20
21	22	23	24	25	26	27	28	29	30
31	32	33	34	35	36	37	38	39	40
41	42	43	44	45	46	47	48	49	50
51	52	53	54	55	56	57	58	59	60
61	62	63	64	65	66	67	68	69	70
71	72	73	74	75	76	77	78	79	80
81	82	83	84	85	86	87	88	89	90
91	92	93	94	95	96	97	98	99	100
101	102	103	104	105	106	107	108	109	110
111	112	113	114	115	116	117	118	119	120

THINK
When you count by tens, each number is ten more.

3, 13, 23, 33, _____, _____, _____, _____, _____, _____, _____, _____

Share and Show

Use a Counting Chart to count by tens.
Write the numbers.

1. Start on 17.

   17, _____, _____, _____, _____, _____, _____, _____, _____

2. Start on 1.

   1, _____, _____, _____, _____, _____, _____, _____, _____

3. Start on 39.

   39, _____, _____, _____, _____, _____, _____, _____, _____

three hundred thirty-eight

© Houghton Mifflin Harcourt Publishing Company

Name _____

MATHEMATICAL PRACTICE ⑤ Use Patterns Use a Counting Chart. Count by tens. Write the numbers.

4. 40, ____, ____, ____, ____, ____, ____, ____

5. 15, ____, ____, ____, ____, ____, ____, ____

6. 28, ____, ____, ____, ____, ____, ____, ____

7. 6, ____, ____, ____, ____, ____, ____, ____

8. 14, ____, ____, ____, ____, ____, ____, ____

9. 32, ____, ____, ____, ____, ____, ____, ____

10. **THINK SMARTER** If you start on 43 and count by tens, what number is after 73 and before 93?

____

11. You say me when you start on 21 and count by tens. I am after 91. I am before 111. What number am I?

____

Problem Solving • Applications  WRITE Math

GO DEEPER Use what you know about a Counting Chart to write the missing numbers.

12.

6		
16		
		28	29

13.

	54	55
	64	
72		

14.

		15
32		

15.

97	98	

16. THINK SMARTER Use a Counting Chart. Count by tens. Match each number on the left to a number that is 10 more.

57 •   • 103

73 •   • 67

77 •   • 87

93 •   • 83

 TAKE HOME ACTIVITY • Write these numbers: 2, 12, 22, 32, 42.
Ask your child to tell you the next 5 numbers.

Count by Tens to 120

COMMON CORE STANDARD—1.NBT.A.1
Extend the counting sequence.

Use a Counting Chart.
Count by tens.
Write the numbers.

1. 1, ____, ____, ____, ____, ____, ____, ____, ____, ____

2. 14, ____, ____, ____, ____, ____, ____, ____, ____, ____

3. 7, ____, ____, ____, ____, ____, ____, ____, ____, ____

4. 29, ____, ____, ____, ____, ____, ____, ____, ____

5. 5, ____, ____, ____, ____, ____, ____, ____, ____, ____

6. 12, ____, ____, ____, ____, ____, ____, ____, ____, ____

7. 26, ____, ____, ____, ____, ____, ____, ____, ____, ____

Problem Solving

Solve.

8. I am after 70.
   I am before 90.
   You say me when you count by tens.
   What number am I? ____

9. **WRITE** Math Use numbers to _____
   explain the pattern you see _____
   when you count forward _____
   by tens. _____

1. Count by tens.
   Write the missing numbers.

   44, 54, 64, _____, _____, 94

2. Use the model. Write to show
   how you make a ten. Then add.

   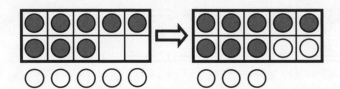

   ___ + ___ + ___

   ___ + ___ = ___

   So, ___ + ___ = ___

3. Write a number sentence
   to complete the related facts.

   9 + 6 = 15   15 − 6 = 9

   6 + 9 = 15   _____

FOR MORE PRACTICE
GO TO THE
Personal Math Trainer

Name _____

Understand Ten and Ones

Essential Question How can you use different ways to write a number as ten and ones?

Common Core **Number and Operations in Base Ten—1.NBT.B.2b**
MATHEMATICAL PRACTICES
MP3, MP5, MP6

Listen and Draw (Real World) (Hands On)

Use 🔲 to model the problem.
Draw the 🔲 to show your work.

Math Talk MATHEMATICAL PRACTICES 5

Use Tools How does your picture show the books Tim has?

🍎 **FOR THE TEACHER** • Read the problem. Tim has 10 books. He gets 2 more books. How many books does Tim have now?

13 is a two-**digit** number.
The 1 in 13 means 1 **ten**.
The 3 in 13 means 3 **ones**.

THINK
10 ones and 3 ones
is the same as
1 ten 3 ones.

__1__ ten __3__ ones

__10__ + __3__

__13__

Share and Show MATH BOARD

Use the model. Write the number
three different ways.

☑ 1.

____ ten ____ ones

___ + ___

___

☑ 2.

____ ten ____ ones

___ + ___

___

Name _____

On Your Own

Use the model. Write the number
three different ways.

3.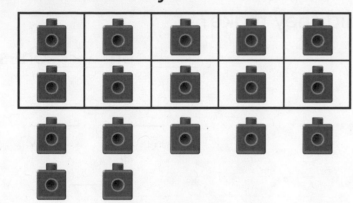

_____ ten _____ ones

_____ + _____

_____

4.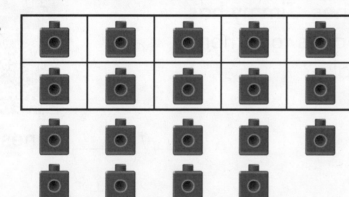

_____ ten _____ ones

_____ + _____

_____

5. 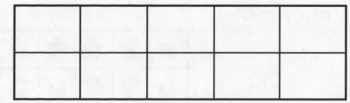 Draw cubes to show the
number. Write the missing numbers.

_____ ten _____ ones

_____ + _____

© Houghton Mifflin Harcourt Publishing Company

Problem Solving • Applications

WRITE Math

Draw cubes to show the number. Write the number three different ways.

6. David has 1 ten and 3 ones. Abby has 6 ones. They put all their tens and ones together. What number did they make?

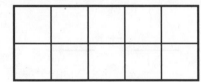

_____ ten _____ ones

_____ + _____

_____

7. THINK SMARTER Karen has 7 ones. Jimmy has 9 ones. They put all their ones together. What number did they make?

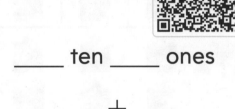

_____ ten _____ ones

_____ + _____

_____

8. THINK SMARTER Does the number match the model?

10 + 5   ○ Yes   ○ No

1 ten 15 ones   ○ Yes   ○ No

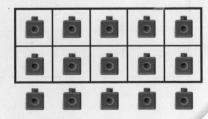

 TAKE HOME ACTIVITY • Show your child one group of 10 small objects and one group of 8 small objects. Ask your child to tell how many tens and ones there are and say the number. Repeat with other numbers from 11 to 19.

Name _____

Understand Ten and Ones

Common Core
COMMON CORE STANDARD—1.NBT.B.2b
Understand place value.

Use the model. Write the number three different ways.

1.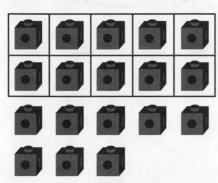

_____ ten _____ ones

___ + ___

___

 Problem Solving Real World

Draw cubes to show the number. Write the number different ways.

Rob has 7 ones. Nick has 5 ones. They put all their ones together. What number did they make?

2.

_____ ten _____ ones

___ + ___

___

3. 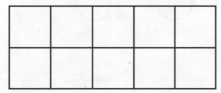 WRITE Math Show twelve in four different ways. Use words, pictures, and numbers.

_____

_____

_____

I. Use the model. Write the
number three different ways.

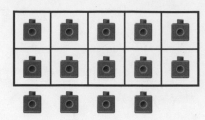

____ ten ____ ones

___ + ___ = ___

___

2. Use the model. Write the
addition sentence. What
number sentence does this
model show?

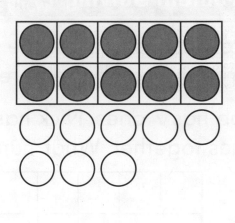

___ + ___ = ___

3. Write two subtraction facts
related to $7 + 5 = 12$.

___ − ___ = ___

___ − ___ = ___

FOR MORE PRACTICE
GO TO THE
Personal Math Trainer

Name _____

Make Ten and Ones

Essential Question How can you show a number as ten and ones?

Common Core Number and Operations in Base Ten—1.NBT.B.2b
MATHEMATICAL PRACTICES
MP2, MP3, MP4, MP6

Listen and Draw *Real World*  **Hands On**

Use to model the problem.
Draw to show your work.

Draw to show the group of ten another way.

 Math Talk

MATHEMATICAL PRACTICES 3

Compare How are the pictures the same? How are the pictures different?

 FOR THE TEACHER • Read the problem. Destiny has 10 cubes. How can she show 1 ten?

© Houghton Mifflin Harcourt Publishing Company

Model and Draw

You can group 10  to make 1 ten.

Draw a quick picture to show 1 ten.

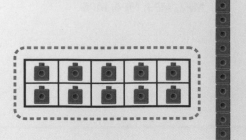

___10___ ones = ___1___ ten

1 ten

Share and Show MATH BOARD

Use . Make groups of ten and ones.
Draw your work. Write how many.

1.

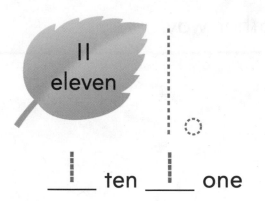

11 eleven

___1___ ten ___1___ one

2.

12 twelve

____ ten ____ ones

☑ 3.

13 thirteen

____ ten ____ ones

☑ 4.

14 fourteen

____ ten ____ ones

Name _____

MATHEMATICAL PRACTICE 6 **Compare** Use ▪. Make groups of
ten and ones. Draw your work. Write how many.

5. 15
    fifteen

    ____ ten ____ ones

6. 16
    sixteen

    ____ ten ____ ones

7. 17
    seventeen

    ____ ten ____ ones

8. 18
    eighteen

    ____ ten ____ ones

9. 19
    nineteen

    ____ ten ____ ones

Problem Solving • Applications  WRITE Math

Solve.

10. **THINK SMARTER** Emily wants to write ten and ones to show 20. What does Emily write?

20
twenty

_____ ten _____ ones

11. **GO DEEPER** Gina thinks of a number that has 7 ones and 1 ten. What is the number? Draw to show your work.

_____

12. Ben drew this picture to show a number. What is the number?

○
○
○
○
○

_____

13. **THINK SMARTER** Circle the numbers that make the sentence true.

There are

| 1 |
| 4 |
| 10 |

tens and

| 1 |
| 4 |
| 10 |

ones in 14.

_____ ten _____ ones

TAKE HOME ACTIVITY • Give your child numbers from 11 to 19. Have your child work with small objects to show a group of ten and a group of ones for each number.

Name _____

Make Ten and Ones

Common Core
COMMON CORE STANDARD—1.NBT.B.2b
Understand place value.

Use 🎲. **Make groups of ten and ones. Draw your work. Write how many.**

1.

14
fourteen

_____ ten _____ ones

2.

12
twelve

_____ ten _____ ones

3.

15
fifteen

_____ ten _____ ones

4.

18
eighteen

_____ ten _____ ones

Problem Solving ⟨Real World⟩

Solve.

5. Tina thinks of a number that has 3 ones and 1 ten. What is the number? _____

6. **Math** Choose a number from 11 to 19. Write the number and number word. Use words and pictures to show how many tens and ones.

Lesson Check (1.NBT.B.2b)

1. How many tens and ones make 17?
   Write the numbers.

17
seventeen

_____ ten _____ ones

Spiral Review (1.OA.A.1, 1.OA.C.6)

2. Use the model. Write the addition sentence. What number sentence does this model show?

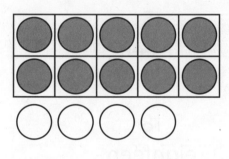

___ + ___ = ___

3. Choose a way to solve. Draw or write to explain. Ben has 17 books. He gives some away. He has 8 left. How many books does he give away?

_____ books

FOR MORE PRACTICE
GO TO THE
Personal Math Trainer

Name _____

Tens

Essential Question How can you model and name groups of ten?

Common Core **Number and Operations in Base Ten—1.NBT.B.2a, 1.NBT.B.2c**
MATHEMATICAL PRACTICES
MP4, MP7, MP8

Listen and Draw Real World

Use to solve the riddle.
Draw and write to show your work.

FOR THE TEACHER • Read the following riddles. I am thinking of a number that is the same as 1 ten and 4 ones. What is my number? I am thinking of a number that is the same as 1 ten and 0 ones. What is my number?

Math Talk
MATHEMATICAL PRACTICES 7
Look for Structure
What did you do to solve the first riddle?

Model and Draw

You can group ones to make tens.

> Draw a quick picture to show the tens.

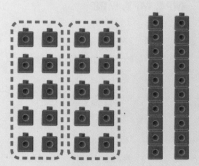

20 ones = __2__ tens __0__ ones

__2__ tens = __20__
twenty

Share and Show · MATH BOARD

Use . Make groups of ten.
Write the tens and ones.

> Draw the tens.
> Count by tens.

1.

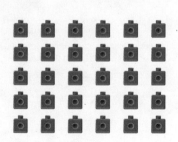

30 ones = ____ tens ____ ones

____ tens = ____
thirty

2.

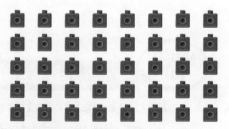

40 ones = ____ tens ____ ones

____ tens = ____
forty

356 three hundred fifty-six

© Houghton Mifflin Harcourt Publishing Company

Name _____

On Your Own

Draw the tens.
Count by tens.

 **MATHEMATICAL PRACTICE (8)** Use Repeated Reasoning

Use . Make groups of ten. Write the tens and ones.

3. 50 ones

____ tens ____ ones

____ tens = ____
fifty

4. 60 ones

____ tens ____ ones

____ tens = ____
sixty

5. 70 ones

____ tens ____ ones

____ tens = ____
seventy

6. 80 ones

____ tens ____ ones

____ tens = ____
eighty

7. 90 ones

____ tens ____ ones

____ tens = ____
ninety

8. **THINK SMARTER** 100 ones

____ tens ____ ones

____ tens = ____
hundred

Name _____

Personal Math Trainer
Online Assessment
and Intervention

Concepts and Skills

Use a Counting Chart.
Count forward. Write
the numbers. (1.NBT.A.1)

1. 63, 64, ____, ____, ____

2. 108, 109, ____, ____, ____

Use a Counting Chart.
Count by tens. Write
the numbers. (1.NBT.A.1)

3. 42, 52, ____, ____, ____

4. 79, 89, ____, ____, ____

5. Use the model. Write the number
three different ways. (1.NBT.B.2b)

____ ten ____ ones

____ + ____

____

Use . Make groups of ten and ones.
Draw your work. Write how many. (1.NBT.B.2b)

6.

15
fifteen

____ ten ____ ones

7. **THINK SMARTER** Choose all the ways that
name the model.
   ○ 60
   ○ 60 tens
   ○ 6 tens 0 ones

Tens

Use . Make groups of ten.
Write the tens and ones.

Common Core   **COMMON CORE STANDARDS—**
1.NBT.B.2a, 1.NBT.B.2c
Understand place value.

I.   90 ones

____ tens = ____ ones   ____ tens = ____
                                                              ninety

....

2.   50 ones

____ tens = ____ ones   ____ tens = ____
                                                              fifty

....

3.   40 ones

____ tens = ____ ones   ____ tens = ____
                                                              forty

Problem Solving Real World

Look at the model. Write the number.

4. What number does the model show?

____

5.   WRITE Math   Draw a quick
picture and write a number
to show thirty.

1. What number does the model show?
Write the number.

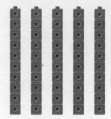

_____ tens = _____

2. What number does the model show?
Write the number.

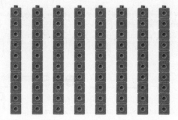

_____ tens = _____

Spiral Review (1.OA.B.3, 1.OA.D.8)

3. Write the missing number.

$$6 + \boxed{} = 13$$

4. What is the sum for $3 + 3 + 4$?

_____

Name _____

Tens and Ones to 50

Essential Question How can you group cubes to show a number as tens and ones?

Common Core
Number and Operations in Base Ten—1.NBT.B.2
MATHEMATICAL PRACTICES
MP4, MP5, MP6

Listen and Draw

Use ▪ to model the number.
Draw ▪ to show your work.

Tens	Ones

Math Talk MATHEMATICAL PRACTICES 4

Model How did you figure out how many tens and ones are in 23?

 **FOR THE TEACHER •** Ask children to use 23 cubes and show them as tens and ones.

Model and Draw

The 2 in 24 means 2 tens.

Tens	Ones

__2__ tens __4__ ones = __24__

The 2 in 42 means 2 ones.

Tens	Ones

__4__ tens __2__ ones = __42__

Share and Show MATH BOARD

Use your MathBoard and 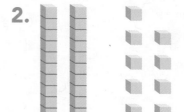 to show the tens and ones. Write the numbers.

1.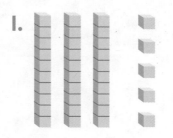

____ tens ____ ones = ____

2.

____ tens ____ ones = ____

✓ 3.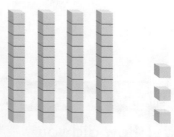

____ tens ____ ones = ____

✓ 4.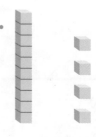

____ ten ____ ones = ____

Name _____

MATHEMATICAL PRACTICE 6 **Make Connections**

Write the numbers.

5.

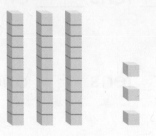

_____ tens _____ ones = _____

6.

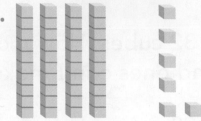

_____ tens _____ ones = _____

7.

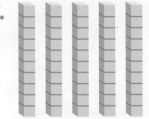

_____ tens _____ ones = _____

8.

_____ tens _____ ones = _____

9. **GO DEEPER** Mary drew tens and ones to show 32.
She made a mistake.
Draw a correct quick picture to show 32.
Write the numbers.

Tens	Ones

Tens	Ones

_____ tens _____ ones = _____

Problem Solving • Applications

 **WRITE** Math

Solve. Write the numbers.

10. I have 46 cubes. How many
tens and ones can I make?

_____ tens _____ ones

11. I have 32 cubes. How many
tens and ones can I make?

_____ tens _____ ones

12. I have 28 cubes. How many
tens and ones can I make?

_____ tens _____ ones

13. **THINK SMARTER** I am a number less
than 50. I have 8 ones and some
tens. What numbers could I be?

_____

Personal Math Trainer

14. **THINK SMARTER +** There are 35 ▪️. Jun says that there
are 3 ones and 5 tens. Rob says that there are
3 tens and 5 ones. Who is correct? Circle the name.

Jun   Rob

How can you draw to show 35?

 TAKE HOME ACTIVITY • Write a two-digit number from 20 to 50,
such as 26. Ask your child to tell which digit names the tens and
which digit names the ones. Repeat with different numbers.

Tens and Ones to 50

Common Core **COMMON CORE STANDARDS—1.NBT.B.2**
Understand place value.

Write the numbers.

1.

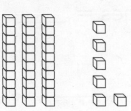

_____ tens _____ ones = _____

2.

_____ tens _____ ones = _____

Problem Solving Real World

Solve. Write the numbers.

3. I have 43 cubes. How many tens and ones can I make?

_____ tens _____ ones

---

4. **WRITE ▶ Math** Write a number from 20 to 50 that has both tens and ones. Use pictures and words to show the tens and ones.

1. What number does the model show?
   Write the numbers.

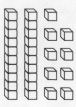

____ tens ____ ones = ____

2. What number does the model show?
   Write the numbers.

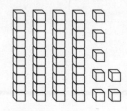

____ tens ____ ones = ____

Spiral Review (1.OA.A.1, 1.OA.C.6)

3. Write the sum.

$$\begin{array}{r} 6 \\ + \underline{3} \\ \end{array}$$

4. Show taking from. Circle the
   part you take from the group.
   Then cross it out.
   Write the difference.

$6 - 4 = \underline{}$

**FOR MORE PRACTICE
GO TO THE
Personal Math Trainer**

Name _____

Tens and Ones to 100

Essential Question How can you show numbers to 100 as tens and ones?

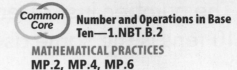

Common Core **Number and Operations in Base Ten—1.NBT.B.2**
MATHEMATICAL PRACTICES
MP.2, MP.4, MP.6

Listen and Draw

Use ▭▭▭▭▭ ▭ to model the number.
Draw a quick picture to show your work.

25

50

52

FOR THE TEACHER • Ask children to use base-ten blocks to show how many tens and ones there are in 25, 50, and 52.

Math Talk MATHEMATICAL PRACTICES ②

Reasoning How did you figure out how many tens and ones are in 52?

Chapter 6

The number just after 99 is 100.
10 tens is the same as I **hundred**.

Draw quick pictures to show 99 and 100.

_9_ tens _9_ ones = _99_ | _10_ tens _0_ ones = _100_

Share and Show MATH BOARD

Use your MathBoard and  to
show the tens and ones. Write the numbers.

1.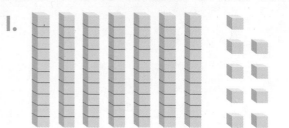

____ tens ____ ones = ____

2.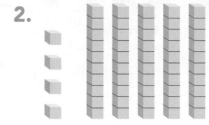

____ tens ____ ones = ____

✓ 3.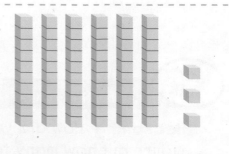

____ tens ____ ones = ____

✓ 4.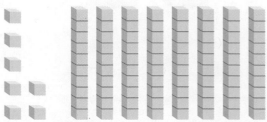

____ tens ____ ones = ____

Name _____

MATHEMATICAL PRACTICE 2 **Reason Quantitatively** Write the numbers.

5.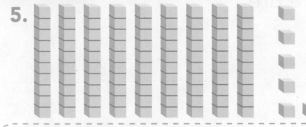

____ tens ____ ones = ____

6.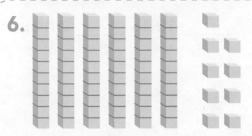

____ tens ____ ones = ____

7.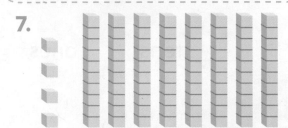

____ tens ____ ones = ____

8.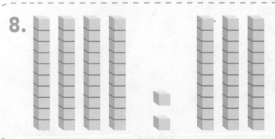

____ tens ____ ones = ____

9.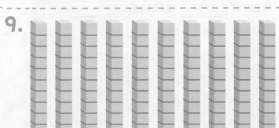

____ tens ____ ones = ____

10. **GO DEEPER** What number is the same as 7 tens and 20 ones?

____

11. **GO DEEPER** What number is the same as 5 tens and 13 ones?

____

Problem Solving • Applications

Draw a quick picture to show the number.
Write how many tens and ones there are.

12. Edna has 82 stamps.

____ tens ____ ones

13. Amy has 79 beads.

____ tens ____ ones

14. **THINK SMARTER** Moe has a
group of 70 red feathers
and 30 brown feathers.

____ tens ____ ones

15. **THINK SMARTER** Read the problem. Write
a number to solve.

I am greater than 14.
I am less than 20.
I have 6 ones.

_____

 TAKE HOME ACTIVITY • Give your child numbers from 50 to 100.
Ask your child to draw a picture to show the tens and the ones in
each number and then write the number.

Name _____

Tens and Ones to 100

COMMON CORE STANDARDS—1.NBT.B.2
Understand place value.

Write the numbers.

1. ____ tens ____ ones = ____

2. ____ tens ____ ones = ____

3. ____ tens ____ ones = ____

Problem Solving Real World

Draw a quick picture to show the number.
Write how many tens and ones there are.

4. Inez has 57 shells.

____ tens ____ ones

5. **WRITE Math** Use words and pictures to show 59 and 95.

Lesson Check (1.NBT.B.2)

1. What number has 10 tens 0 ones?

_____

2. What number does
the model show?
Write the numbers.

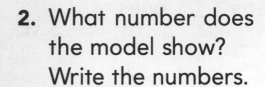

_____ tens _____ ones = _____

Spiral Review (1.OA.B.3, 1.OA.C.5)

3. Barry knows that $6 + 5 = 11$.
What other addition fact does
he know? Write the new fact.

_____ + _____ = _____

4. Count on to solve $2 + 6$.
Write the sum.

$2 + 6 = $ _____

FOR MORE PRACTICE
GO TO THE
Personal Math Trainer

Problem Solving • Show Numbers in Different Ways

Essential Question How can making a model help you show a number in different ways?

Common Core Number and Operations in Base Ten—1.NBT.B.2a, 1.NBT.B.3
MATHEMATICAL PRACTICES
MP1, MP4, MP6, MP7

Gary and Jill both want 23 stickers for a class project. There are 3 sheets of 10 stickers and 30 single stickers on the table. How could Gary and Jill each take 23 stickers?

 Unlock the Problem Real World

What do I need to find?

_____

___**two**___ different ways to make a number

What information do I need to use?

The number is __23__.

Show how to solve the problem.

Gary	
Tens	Ones

23

Jill	
Tens	Ones

23

© Houghton Mifflin Harcourt Publishing Company

HOME CONNECTION • Showing the number with base-ten blocks helps your child explore different ways to combine tens and ones.

Try Another Problem

Use ▭▭▭▭▭ ▪ to show the number two different ways. Draw both ways.

1. 46

Tens	Ones

Tens	Ones

____ ◯ ____

2. 71

Tens	Ones

Tens	Ones

____ ◯ ____

3. 65

Tens	Ones

Tens	Ones

____ ◯ ____

Math Talk

MATHEMATICAL PRACTICES 6

Look at Exercise 3. **Explain** why both ways show 65.

Name _____

Use ▭▭▭▭ ▫ to show the number
two different ways. Draw both ways.

☑ **4.**   **59**

Tens	Ones

Tens	Ones

____ ◯ ____

☑ **5.**   **34**

Tens	Ones

Tens	Ones

____ ◯ ____

6. THINK SMARTER Show 31 three ways.

Tens	Ones

Tens	Ones

Tens	Ones

____ ◯ ____ ◯ ____

On Your Own   **WRITE** Math

Write a number sentence to solve. Draw to explain.

7. **MATHEMATICAL PRACTICE ④** **Write an Equation**
Felix invites 15 friends to his
party. Some friends are girls.
8 friends are boys. How many
friends are girls?

___ ◯ ___ ◯ ___
girls

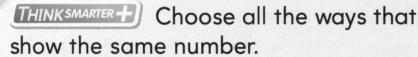 **GO DEEPER** Solve. Write the numbers.

8. I am a number less than 35.
I have 3 tens and some ones.
What numbers can I be? _____

Personal Math Trainer

9. **THINK SMARTER +** Choose all the ways that
show the same number.

 TAKE HOME ACTIVITY • Have your child draw quick pictures
to show the number 56 two ways.

376 three hundred seventy-six

Problem Solving • Show Numbers in Different Ways

COMMON CORE STANDARDS—1.NBT.B.2a,
1.NBT.B.3 *Understand place value.*

Use ▭▭ ▢ to show the number
two different ways. Draw both ways.

1.   62

Tens	Ones

_____ ◯

Tens	Ones

_____

2.   38

Tens	Ones

_____ ◯

Tens	Ones

_____

3. **WRITE** Math Draw to show 55 three different ways.

Lesson Check (1.NBT.B.2a, 1.NBT.B.3)

1. What number does each model show? Write the numbers.

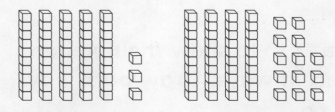

_____ = _____

2. What number does the model show? Write the number.

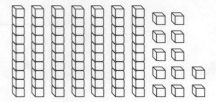

_____

Spiral Review (1.OA.C.6)

3. Subtract to solve. Then add to check your answer.

$12 - 4 = \boxed{}$

$\boxed{} + 4 = \boxed{}$

4. Write two ways to make 15.

$15 = \underline{} + \underline{}$ $15 = \underline{} + \underline{}$

FOR MORE PRACTICE
GO TO THE
Personal Math Trainer

Name _____

Model, Read, and Write Numbers from 100 to 110

Essential Question How can you model, read, and write numbers from 100 to 110?

Common Core Number and Operations in Base Ten—1.NBT.A.1
MATHEMATICAL PRACTICES
MP.4, MP.5, MP.7

Listen and Draw Real World

Use .
Circle a number to answer the question.

1	2	3	4	5	6	7	8	9	10
11	12	13	14	15	16	17	18	19	20
21	22	23	24	25	26	27	28	29	30
31	32	33	34	35	36	37	38	39	40
41	42	43	44	45	46	47	48	49	50
51	52	53	54	55	56	57	58	59	60
61	62	63	64	65	66	67	68	69	70
71	72	73	74	75	76	77	78	79	80
81	82	83	84	85	86	87	88	89	90
91	92	93	94	95	96	97	98	99	100

Math Talk
MATHEMATICAL PRACTICES 7

Look for Structure
Why is 100 to the right of 99 on the hundred chart? Why is 100 below 90?

FOR THE TEACHER • Have children locate each number on the hundred chart. What number is the same as 30 ones? What number is the same as 10 tens? What number is the same as 8 tens 7 ones? What number has 1 more one than 52? What number has 1 more ten than 65?

three hundred seventy-nine **379**

Model and Draw

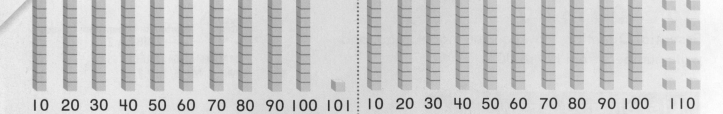

10 20 30 40 50 60 70 80 90 100 101   10 20 30 40 50 60 70 80 90 100 110

10 tens and 1 more = **101**   10 tens and 10 more = **110**

Share and Show MATH BOARD

Use  to model the number.
Write the number.

> **REMEMBER**
> 10 tens = 100

1. 10 tens and 1 more

_____

2. 10 tens and 2 more

_____

3. 10 tens and 3 more

_____

4. 10 tens and 4 more

_____

☑5. 10 tens and 5 more

_____

☑6. 10 tens and 6 more

_____

Name _____

On Your Own

MATHEMATICAL PRACTICE 4 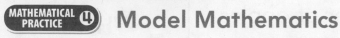 **Model Mathematics**

Use  to model the number.
Write the number.

7. 10 tens and
7 more

_____

8. 10 tens and
8 more

_____

9. 10 tens and
9 more

_____

10. 10 tens and 10 more

_____

11. **THINK SMARTER** 11 tens

_____

Write the number.

12.

_____

13.

_____

14.

_____

15.

_____

Problem Solving • Applications    Math

GO DEEPER Solve to find the number of apples.

THINK
 = 1 apple

= 10 apples

16.

There are _____ apples.

17.

There are _____ apples.

18.

There are _____ apples.

19. **THINK SMARTER** What number does the model show?

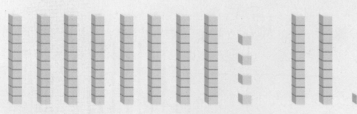

_____

 TAKE HOME ACTIVITY • Give your child a group of 100 to 110 small objects. Ask him or her to make as many groups of ten as possible, then tell you the total number of small objects.

382 three hundred eighty-two

Model, Read, and Write Numbers from 100 to 110

Common Core   COMMON CORE STANDARDS—1.NBT.A.1
Understand place value.

Use to show the number.
Write the number.

1. 10 tens and
6 more

_____

2. 10 tens and
1 more

_____

3. 10 tens and
10 more

_____

Problem Solving Real World

4. Solve to find the number of pens.

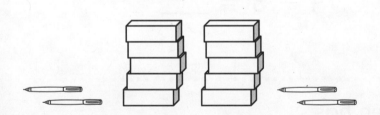

THINK

✐ = 1 pen

▭ = 10 pens

There are _____ pens.

5. WRITE Math   Choose a number from 101 to 110. Write it.
Draw a picture to show it as 10 tens and more.

1. What number does the model show?
Write the number.

_____

2. Show taking from. Circle the
part you take from the group.
Then cross it out.
Write the difference.

$4 - 3 =$ _____

3. Use the model to solve. Ken has
8 toy trains. Ron has 3 toy trains.
How many fewer toy trains does
Ron have than Ken?

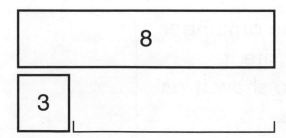

_____

FOR MORE PRACTICE
GO TO THE
Personal Math Trainer

Name _____

Model, Read, and Write Numbers from 110 to 120

Essential Question How can you model, read, and write numbers from 110 to 120?

Common Core Number and Operations in Base Ten—1.NBT.A.1
MATHEMATICAL PRACTICES
MP2, MP4, MP6

Listen and Draw (Real World)

How many shells are there?

There are _____ shells.

Math Talk

MATHEMATICAL PRACTICES 2

Reasoning How did you decide how many shells there are? Explain.

FOR THE TEACHER • The picture shows the shells that Heidi has collected. How many shells does Heidi have?

11 tens is 110.

12 tens is 120.

110

120

Share and Show MATH BOARD

Use  to model the number.
Write the number.

1.

111

2.

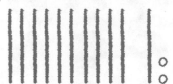

_____

✓**3.**

_____

✓**4.**

_____

Name _____

 **Model Mathematics**

Use  to model the number.
Write the number.

5.

_____

6.

_____

7.

_____

8.

_____

9.

_____

10.

_____

THINK SMARTER Write the number.

11.

_____

12.

_____

13.

_____

Problem Solving • Applications   WRITE Math

GO DEEPER   **Choose a way to solve.**
Draw or write to explain.

14. Joe collects toy cars. He can make
11 groups of 10 toy cars.
How many toy cars
does Joe have?

_____ toy cars

15. Cindy collects buttons. She can
make 11 groups of 10 buttons
and one more group of 7 buttons.
How many buttons
does Cindy have?

_____ buttons

16. Lee collects marbles. He can make
11 groups of 10 marbles
and has 2 marbles left
over. How many marbles
does Lee have?

_____ marbles

17. THINK SMARTER   Finish the drawing to show 119.

Write to explain.

_____

TAKE HOME ACTIVITY • Give your child a group of 100 to
120 small objects. Ask him or her to make as many groups of ten
as possible and then tell you the total number of small objects.

Model, Read, and Write Numbers from 110 to 120

COMMON CORE STANDARDS—1.NBT.A.1
Extend the counting sequence.

Use to model the number.
Write the number.

1.

____

2.

____

3.

|||||||||

____

Problem Solving Real World

Choose a way to solve. Draw or write to explain.

4. Dave collects rocks. He makes 12 groups of 10 rocks and has none left over. How many rocks does Dave have?

____ rocks

5. **WRITE** Math Choose a number from 111 to 120. Write the number. Draw a picture to show it as tens and ones.

1. What number does the model show?
Write the number.

_____

....

Spiral Review (1.OA.C.6)

2. Show how to make a ten to solve 13 − 7.
Write the number sentence.

Step 1	Step 2

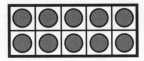 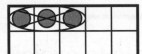 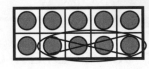 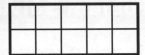

$$\underline{} - \underline{} \quad \underline{}$$

$$\underline{} - \underline{} = \underline{}$$

So, 13 − 7 = ____.

....

3. What is the difference?
Write the number.

$$\begin{array}{r} 9 \\ -\ 4 \\ \hline \end{array}$$

**FOR MORE PRACTICE
GO TO THE
Personal Math Trainer**

✓ Chapter 6 Review/Test

Personal Math Trainer
Online Assessment
and Intervention

1. Felix counts 46 cubes. Then he counts forward some more cubes. Write the numbers.

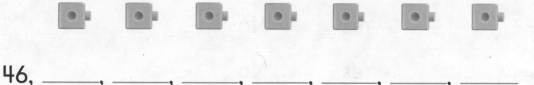

46, _____, _____, _____, _____, _____, _____, _____

2. Count by tens. Match each number on the left to a number that is 10 more.

35 •   • 69

49 •   • 59

59 •   • 75

65 •   • 45

57 •   • 67

3. Does the number match the model?
Choose Yes or No.

10 + 10	○ Yes	○ No
1 ten 4 ones	○ Yes	○ No
1 ten 5 ones	○ Yes	○ No
10 + 5	○ Yes	○ No

4. Circle the numbers that make the sentence true.

There are
| 1 |
| 2 |
| 10 |
tens and
| 1 |
| 2 |
| 10 |
ones in 12.

5. Choose all the ways that
name the model.

○ 3 ones

○ 3 tens

○ 3 tens 0 ones

○ 30

6. There are 42 . Lisa says that there are
4 tens and 2 ones. Elena says there are 2 tens
and 4 ones. Who is correct? Circle the name.

Lisa   Elena

How can you draw to show 42?

---

7. Read the problem. Write a number to solve.

I am greater than 27.
I am less than 30.
I have 9 ones.

8. **THINK** SMARTER **+**   Choose all the ways that
show the same number.

Personal Math Trainer

○

○   ○

○   ○

9. What number does the model show?

|||||||||| o o
            o o
            o o
            o o
            o o

_____

---

10. GO DEEPER   Finish the drawing to show 118.

|||||||||| o
            o
            o
            o
            o o

Write to explain.

---

11. Count the 🌰. Write the numbers.

How many tens? _____ tens

How many 🌰? _____

_____

---

12. Draw a quick picture to show 54 in two ways. Then write the number of tens and ones in each picture.

_____ tens _____ ones   _____ tens _____ ones